I0484380

-ABSTRACT-

In the course of researching basic rocket fuel combustion technology; a diagram was obtained from the internet with a related formula that did not equate as originally structured.

Therefore, a new understanding for the above formula was thought to be required for equaling the intended mix ratio (r) of the propellants with their injection angles and where they have an axial flow of a zero tan along the combustion chamber axis.

The work in this paper culminates in two restructured formulas utilizing a new factor called the Velocity Ratio (Vr). It was developed with basic geometry, trigonometry and some algebraic experimentation that allowed the results of the new formulas to equate with and correspond with the concept detailed in the referenced diagram for bi-propellant combustion in impingement injection fueling systems.

This Velocity Ratio (Vr), is the quotient of the oxidant velocity divided by the fuel velocity.

-NOMENCLATURE-

C effective exhaust velocity FT/SEC

I_{sp} specific impulse

g gravitational constant 32.2 FT/SEC/SEC

q mass flow lbs/SEC

F thrust--force lbs

W_t total propellant weight lbs.

W_f fuel propellant weight lbs.

W_o oxidant propellant weight lbs.

m_o oxidant mass flow lbs/sec

m_f fuel mass flow lbs/sec

θ impingement angle   degrees

X_A remaining degrees $(180 - \theta)$

r mix ratio

γ_f fuel injection angle degrees

γ_o oxidant injection angle degrees

-NOMENCLATURE-

C_c coefficient of jet contraction

A_d nozzle tip diameter area in2

A_D feed line diameter area in2

C_d orifice discharge coefficient

ΔP change in pressure p.s.i

ρ propellant density slug/ft3 (combustion) , lbs/ft3

x length between impingement Pt & injection plane

V_o oxidant velocity   ft/sec

V_f fuel velocity   ft/sec

V_r velocity ratio (V_o/V_f)

$\tan \theta$ axial flow of zero

According to the textbook "Rocket Propulsion Elements" authored by George P. Sutton & Oscar Biblarz; "a complete theory relating injector design parameters to rocket performance and combustion phenomena has yet to been devised and…. has been largely empirical."[1]. Conceptually, they have a reasonable approach; however, their equations fail to equate and must be restructured to verify their concept for rocket fuel combustion since their formulas were never fully examined to be questioned for verification.

This approach for designing impingement injectors alters nearly 60 years of propulsion theory by creating a sound mathematical structure that designs impingement injectors with a pair of restructured formulas that were created through algebraic experimentation to equate with the intended mix ratio (r) while having an axial flow of the propellants that is equal to a tan of zero when using a new ratio that will be defined as the Velocity Ratio (Vr) which is the oxidant velocity is divided by the velocity of fuel.

A primary reason the Sutton & Biblarz formulas do not equate is because, the velocities of the propellant differ from their angles of injection due to their change in pressure with their mass flow along and their densities therefore, requiring further consideration to have their basic concept restructured with basic geometry and mathematics.

This paper provides the means to accurately design impingement injectors with an emphasis on constructing the injection angles that equate within a new pair of formulas for optimal combustion of propellants using the Ratio of Velocity (Vr) that confirms them for the intended concept created by Sutton & Biblarz.

Balanced Injector Orientation
for Propellant Combustion
Elmer A. Beck

The manner to accomplish and put forth this concept of a Velocity Ratio (V_r) begins with designing the impingement injectors with the details that construct the injection angles using the mix ratio along with basic geometry and mathematics then determine the propellant velocities before the introduction and the creation of the Velocity Ratio (V_r).

The first item to calculate is the mass flow of the oxidant and fuel using a mix ratio (r) of 3.5 for a thrust (F) of 74,200 lbs with basic formulas from web sites such as 1) braeunig.us 2) risacher.org along with charts from other web sites for Specific Impulse (I_{sp}) & mix ratios for fueling.

Prior to calculating the mass flows, the effective exhaust velocity (C) is determined.

$$C = I_{sp}\,(g) \qquad\qquad\qquad (\text{EQ. 1})$$

g: Gravitational Constant, 32.2 ft /sec/ sec

$$C = 363(32.2)$$

$$= 11,688.6 \ \text{ft} / \text{sec}$$

Calculating the mass flows.

$$q = F/C \qquad \text{(EQ. 2)}$$

q : Mass Flow

$$q = 74{,}200 / 11{,}688.6$$

$$= 6.34806564 \text{ lbs./sec}$$

A self-check will be completed to verify that the Specific Impulse (Isp) equals the thrust divided by the product of the mass flow and the gravitational constant (g).

$$Isp = \frac{F}{qp} \qquad \text{(EQ. 3)}$$

$$Isp = \frac{74{,}200}{6.34806564(32.2)} = \frac{74{,}200}{204.407714} = 363$$

This quotient correlates with the variables to calculate the Specific Impulse (Isp).

Balanced Injector Orientation
for Propellant Combustion
Elmer A. Beck

The next items to determine are the oxidant and fuel mass flows with formulas from riacher.org then their values are used for calculating their impingement angle with a total mass flow of 6.34806564 lbs/sec in a 3.5 mix ratio.

$$W_o = W_t (r)/ (r +1) \qquad\qquad (EQ.\ 4)$$

$$W_o = 6.34806564\ (3.5) / (3.5 + 1)$$

$$= 22.2182297 / 4.5$$

$$= 4.9373844\ \text{lbs/sec}$$

$$W_f = W_t/ (r+1) \qquad\qquad (EQ.\ 5)$$

$$W_f = 6.34806564/ (3.5+1)$$

$$= 6.34806564/ 4.5$$

$$= 1.41068125\ \text{lbs/sec}$$

Self-check to verify the results

$$W_t = W_o + W_f \qquad\qquad (EQ.\ 6)$$

$$W_t = 4.93738 + 1.41068$$
$$= 6.34806564$$

The mass flows equate correctly.

Balanced Injector Orientation
for Propellant Combustion
Elmer A. Beck

8/46

The following diagram [2] depicts the basic concept of impingement injection and was the guide to explain how to adhere to this concept while restructuring the two typical formulas related to this diagram that are a mathematical confirmation for optimal propellant combustion using these impingement injection systems.

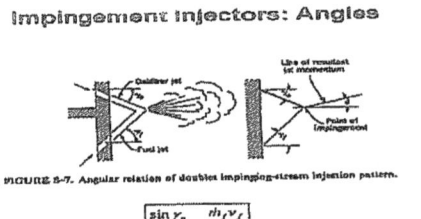

Impingement Injectors: Angles

FIGURE 5-7. Angular relation of double impinging-stream injection pattern.

$$\frac{\sin \gamma_o}{\sin \gamma_f} = \frac{\dot{m}_f v_f}{\dot{m}_o v_o}$$

From Sutton

FIG. #1 [2]

In order to determine the propellant injection angles, the angle of propellant impingement must be calculated.

This impingement orientation was shown in Figure #1 [2] which is the basis for this paper's goal to confirm with mathematically with a couple of new formulas using a velocity ratio to prove the concept depicted in Figure #1.

Although, the propellant streams in FIG. #1 resembles a right triangle; this is not the case and it will be proven by mathematics with trigonometry and geometric construction.

After determining the propellants' mass flow, the angle of propellant impingement is calculated by dividing the oxidant mass flow by the fuel mass flow which equals the co-tangent for the impingement angle while also being the intended mix ratio for the propellants

The result of the above division using the mass flow parameters will be drawn in Figure #2 and is similar to the flows of propellant in Figure #1 by Sutton & Biblarz.

$$\cot \theta = \frac{m_o}{m_f} = \frac{4.9374}{1.41068} \qquad \text{(EQ. 7)}$$

$$\cot \theta = 3.500014178$$

$$\theta = 74.00547 = 74° = \text{Impingement Angle}$$

NOTE: The co-tangent value equals the Mix Ratio

Therefore, the quotient of the propellant mass flow division yields the impingement angle (θ) by converting the co-tangent value which is illustrated in Figure #2. If the mass flow of the propellants were reversed, the resulting quotient would be the reciprocal of the co-tangent value.

The next step to complete in determining the propellant injection angles is to draw a vertical line at the impingement point which is equal to a dimension of 180 degrees according to basic geometry and is illustrated in Figure #3._

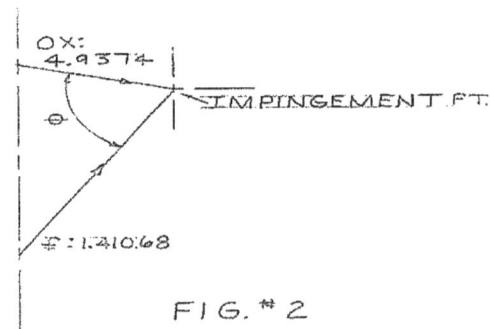

FIG. # 2

Balanced Injector Orientation
for Propellant Combustion
Elmer A. Beck

11/46

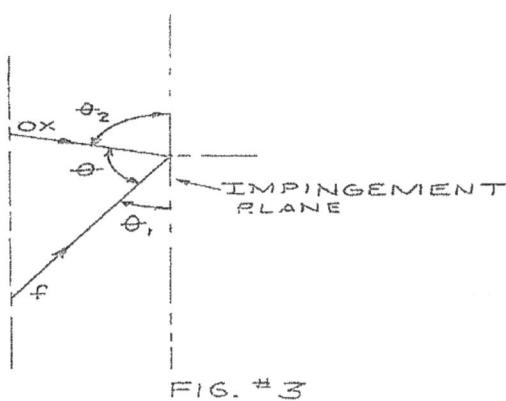

FIG. #3

The following pertains to detailing the intermediate steps to calculate the impingement angles (θ_1), (θ_2) as depicted in Figure #3.

The Impingement Angle (θ) was calculated to be 74 Degrees and will be subtracted from 180 Degrees which is the dimension of a straight line; resulting with a remainder of 106 Degrees that will be divided by the value of the Mix Ratio (r), since that is the value for propellant combustion.

Balanced Injector Orientation
for Propellant Combustion
Elmer A. Beck

12/46

$$X_A = 180 - \text{Imp. Angle } (\theta) \qquad \text{(EQ. 8)}$$

$$X_A = 180 - 74$$

$$= 106 \text{ Degrees}$$

$$\theta = X_A / r \qquad \text{(EQ. 9)}$$

$$\theta = 106 / 3.5$$

$$= 30.286 \text{ Degrees}$$

The above division yields a quotient that will be added to the impingement angle (θ) for a sum that is subtracted from 180 Degrees to yield another angle along the plane of impingement that is 180 Degrees.

$$\theta = 180 - (\text{Impg.}\theta + \theta_1) \qquad \text{(EQ. 10)}$$

$$\theta = 180 - (74 + 30.286)$$

$$= 180 - 75.714$$

$$\theta = 75.714 \text{ Degrees}$$

In order to assign angle values in Figure #3 along the impingement plane near the impingement point to design the propellant injection angles; it is worth considering that in the Mix Ratio (r), the quantity of oxidant is greater than fuel and the impingement angles should reflect this factor.

Then, the Angle (θ_1) of lesser value will be assigned between the impingement plane and the fuel stream while the larger Angle (θ_2) will be assigned between the plane of impingement and the oxidant stream.

A quick self-check of the impingement angles will determine if they are equal to 180 Degrees because that is the dimension of a straight line which is the impingement plane in this propellant injection design.

$$180 = \theta_1 + \theta_2 + \theta$$

$$180 = 30.286 + 75.714 + 74$$

$$180 = 106 + 74$$

$$180 = 180$$

This self-check verifies the calculated angles for the impingement of the propellants.

The determination of the injection angles for the propellants are concluded by creating right triangles whose hypotenuse are the propellant streams as illustrated in the next couple of figures.

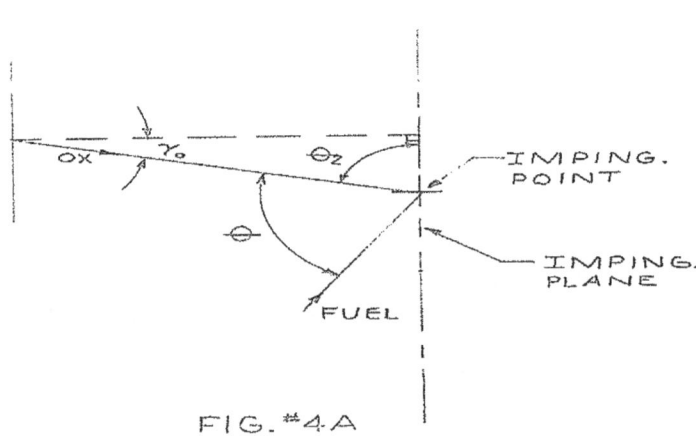

FIG. #4A

Since Right Triangles consist of 180 Degrees and one of those angles is 90 Degrees, as indicated with a small square; the Injection Angles (γ) would be the remainder of 90 Degrees minus the impingement angle near the impingement point.

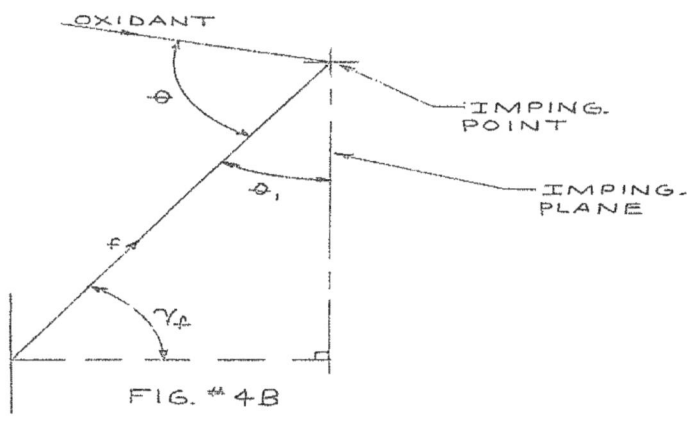

FIG. # 4B

$$\theta_o = 90 - \theta_2 \qquad\qquad \theta_f = 90 - \theta_1$$

$$= 90 - 75.71 \qquad\qquad = 90 - 30.28$$

$$\theta_o = 14.28 \text{ Deg.} \qquad\qquad \theta_f = 59.71 \text{ Deg.}$$

To confirm the injection angles, a self-check will be completed by dividing the sin of the fuel injection angle by the sin of the oxidant injection angle whose quotient should equal the Mix Ratio of 3.5 of the impingement injection.

Since the fuel injection angle is greater than the oxidant injection angle, their manner of division is more logical for equating with the Mix Ratio for combustion.

$$r = \frac{\sin \gamma_f}{\sin \gamma_o} \qquad \text{(EQ. 11)}$$

$$r = \frac{\sin 59.71}{\sin 14.28}$$

$$= 3.50069$$

$$r \cong 3.5$$

The results of the above calculations were carried out to two decimal places for verifying their equivalency for greater accuracy with the intended mix ratio of 3.5.

These angles of injection are also the parameters used to determine the change in pressure of the propellants with their density along with their nozzle orifice discharge coefficient that calculates the propellant injection velocities.

To determine the propellant velocities, the value of the Orifice Discharge Coefficient (*Cd*) must be calculated with the Coefficient of Jet Contraction (*Cc*) which is the ratio between the propellant feed-line area and the nozzle tip area.

For approximating a nozzle tip diameter, the propellant density (ρ) will be a measure of stress acting on a plane that would be the nozzle tip diameter and begins with calculating the force of the propellants.

The force (*F*) is the product of the propellant mass flow and the gravity constant (*g*) which is then divided by the Specific Impulse (*Isp*).

$$FORCE = \frac{qg}{Isp} \qquad \text{(EQ. 12)}$$

The previous formula is based upon having the value of Specific Impulse equaling the force divided by the product of the mass flow and the gravity constant as according to standard rocketry formulas.

$$Isp = \frac{F}{qg} \qquad \text{(EQ. 3)}$$

For the purpose of calculating the nozzle tip diameter, the area equals the stress of the propellant density dividing the force or, mass flow and is used to calculate the orifice discharge coefficient (*Cd*).

Using standard gas density values for determining the nozzle diameter yields fairly large dimensions which may not reflect the nature of mass flow in this example of rocket fuel combustion.

From a text on Fluid Mechanics, the measurement of mass usually involves comparisons of gravitational attraction exerted upon the body of matter in question [3].

Area = Force / Stress

Area = Mass Flow / Propellant Density 'ρ' (EQ. 13)

PROPELLANT DENSITY   [4]

Hydrogen   0.000165 slug / ft3
Oxygen   0.00262   slug / ft3

Therefore, since slug/ft3 is often associated with rocket propellants it will be utilized in the following calculations to approximate this nozzle tip diameter.

$$F = qg / Isp \qquad \text{(EQ. 12)}$$

$$F_{ox} = \frac{4.9374\ (32.2)}{363} = 0.437973 \text{ lbs.}$$

Balanced Injector Orientation
for Propellant Combustion
Elmer A. Beck

$$F_f = \frac{1.41068\,(32.2)}{363} = 0.1251347 \text{ lbs.} \qquad \text{(EQ. 12)}$$

$$\text{Area} = \frac{\rho}{\text{FORCE}} \qquad \text{(EQ. 14)}$$

OXIDANT	FUEL

$$A = \frac{0.000262}{0.437973} \qquad\qquad A = \frac{0.000165}{0.12513}$$

$$= 0.005982 \text{ in2} \qquad\qquad = 0.00131856 \text{ in2}$$

$$d = \sqrt{\frac{4\,A}{\pi}}$$

$$d_{ox} = \sqrt{\frac{(4)\,0.005982}{\pi}} = 0.0727267 \text{ in } (3/32'')$$

$$d_f = \sqrt{\frac{(4)\,0.00131856}{\pi}} = 0.040974 \text{ in } (1/16")$$

The NASA Technical Drawing [5] has a feed-line diameter of ¾ inch and will be the reference dimension to calculate the Coefficient of Jet Contraction (Cc); where the area of the nozzle tip is divided by the area of the feed-line as described by Hunter Rouse's textbook on fluid mechanics [6].

$$Cc = \frac{Ad}{AD} \quad [6] \qquad (EQ.\ 15)$$

$$\text{Feed Line/AD} = \sqrt{\frac{\pi}{4}\,0.75^2} = 0.4417865 \text{ in2}$$

$$\text{Oxidant/ Ad} = \sqrt{\frac{\pi}{4}\,0.09375^2} = 0.006029 \text{ in2}$$

$$\text{Fuel / Ad} = \sqrt{\frac{\pi}{4}\,0.0625^2} = 0.00\,306796 \text{ in2}$$

Calculating Coefficient of Jet Contraction (*Cc*)

Oxidant Cc:

$$Cc = Ad / AD$$

$$Cc = 0.006029 / 0.4417865$$

$$= 0.015625$$

Fuel Cc:

$$Cc = Ad / AD$$

$$Cc = 0.00306796 / 0.4417865$$

$$= 0.0069444$$

Calculating the Orifice Discharge Coefficient (*Cd*) with the formula from [7] will be utilized with the change in pressure to determine the velocities of the propellants.

$$Cd = \frac{Cc}{\sqrt{1 - Cc\text{^}2\ (d/D)\text{exp}4}} \qquad [7] \qquad (EQ.\ 16)$$

Balanced Injector Orientation
for Propellant Combustion
Elmer A. Beck

Oxidant Cd

$$Cd = \frac{0.015625}{\sqrt{1 - 0.015625^2 \ (0.09375/0.75) \ exp4}}$$

$$= 0.015625$$

Fuel Cd

$$Cd = \frac{0.006944}{\sqrt{1 - 0.006944^2 \ (0.0625/0.75) \ exp4}}$$

$$= 0.006944$$

The other parameter to calculate for determining the propellant velocities is the change in pressure from the nozzle orifice to the impingement point of the propellants.

Using Fig. 8-7 [2], the horizontal distance between the propellant injection plane and the impingement plane is approximated by hand sketching from the impingement point left to the injection plane then measured in scale which is best done on ¼ inch grid paper.

After sketching the injection streams from the point of impingement, draw a vertical line where the injection points of entry are connected by a vertical line at the ¼ inch grid intersection points.

Then draw a perpendicular line from the impingement point left to the vertical line for the plane of propellant injection which will serve as the distance between these points in calculating the change in pressure with a standard formula from fluid mechanics and is stipulated as:

$$\Delta P = g\, \rho\, \tan\theta\, x \qquad\qquad (EQ.\ 17)$$

Prior to calculating the change in pressure, keep in mind there are (2) propellant densities due to the phases of gas and liquid where each being compared for velocity and mix ratio accuracy in the forthcoming calculations.

Gas Density	Liquid Density
OX: 27.22	OX: 0.089212
FUEL: 1.88	FUEL: 0.005611

When calculating the change in pressure, take the gravitational constant (g) and multiply it with the propellant density then multiply this product with each propellant injection tan (θ) times the prior stated perpendicular line.

The following standard equation from fluid mechanics is to determine the change in pressure for the (2) phases of each propellant.

$$\Delta P = g\, \rho \tan\theta\, x \qquad\qquad \text{(EQ. 17)}$$

OXIDANT (GAS)

$\Delta P = 32.174(27.22)\tan 14.28(16)$
$\quad = 3566.5126$

OXIDANT (LIQUID)

$\Delta P = 32.174(0.089212)\tan 14.28(16)$
$\quad = 11.68904$

FUEL (GAS)

$\Delta P = 32.174(1.88)\tan 59.72(16)$
$\quad = 1657.50875$

FUEL (LIQUIUD)

$\Delta P = 32.174(0.005611)\tan 59.72(16)$
$\quad = 4.946958$

Balanced Injector Orientation
for Propellant Combustion
Elmer A. Beck

Determining the propellant velocities with the value for Orifice of Discharge Coefficient (*Cd*) and change in pressure with the following formula from the web site risacher.org in the (2) phases of densities for the propellants.

$$v = Cd \sqrt{2g \, (\Delta P / \rho)} \quad [7] \qquad (EQ. \ 18)$$

GAS PHASE

$$Vo = 0.01562 \sqrt{(2)32.174 \, (3566.5126/27.22)}$$
$$= 1.434255 \text{ ft/sec}$$

$$Vf = 0.006944 \sqrt{(2)32.174 \, (1657.50875/1.88)}$$
$$= 1.6540599 \text{ ft/sec}$$

LIQUID PHASE

$$Vo = 0.015611 \sqrt{(2)32.174 \, (11.689/0.089212)}$$
$$= 1.433429 \text{ ft/sec}$$

$$Vf = 0.006944 \sqrt{(2)32.174 \, (4.946958/0.005611)}$$
$$= 1.653964 \text{ ft/sec}$$

The next group of calculations will consider the propellant mass flow multiplied with their velocities to ascertain if their results are equal to the intended mix ratio as implied with the Sutton & Biblarz concept; where their mass flow times velocity of the oxidant is divided by the mass flow of the fuel times it's velocity is to equal the mix ratio as with the division of the sins for the propellant injection angles with the sin γ_f being divided by the sin γ_o.

With the means to determine the propellant velocities, the velocity ratio (V_r) can be defined as the quotient of the oxidant velocity divided by the fuel velocity.

$$V_r = V_o / V_f \qquad \text{(EQ. 19)}$$

This velocity ratio will be utilized in a pair of algebraic equalization statements that create a mathematical footing to confirm the intended design criteria for the propellant injectors with the intended mix ratio and proving that the axial flow of the propellants is along the axis of the combustion chamber which is equal to a tan of zero.

According to FIG. 8-7 [2], the division of the injection angle sins equals the division of the propellants' mass flow times their velocity which is not the case. Therefore, these formulas must be restructured if the concept is to be accepted as a valid mathematical structure for propellant impingement which were accomplishment through algebraic experimentation.

When multiplying the quotient of the propellants' mass flow times their respective velocities with the velocity ratio (V_r), yields a value that is less than intended mix ratio, (r)

$$r = \left[\frac{m_o \, v_o}{m_f \, v_f} \right] (v_r) \qquad \text{(EQ. 20)}$$

However, when dividing the mass flow quotient by the velocity ratio (v_r); the result is more equal to the mix ratio as intended for the propellants in use.

This equation should be equal to the quotient dividing the sin for the fuel injection angle by the sin of the oxidant injection angle which does equal the intended mix ratio.

Therefore, the correct formula to calculate impingement angles with a solid mathematical foundation is as follows and do pertain to the two phases of propellants.

$$r = \left[\frac{m_o \, v_o}{m_f \, v_f} \right] \div v_r \qquad \text{(EQ. 21)}$$

The following calculations illustrate how dividing by the Velocity Ratio (v_r) results in a value that is equal to the intended mix ratio.

Prior to confirming the mix ratio with the correct equation, the velocity ratio will be calculated.

$$v_r = v_o \, / \, v_f \qquad \text{(EQ. 19)}$$

$$= 1.4343 \, / \, 1.6541 \quad = 0.86711807$$

EQUATION (20)

$$r = \left[\frac{4.9374\,(1.4343)}{1.41068(1.6541)} \right] (0.86711)$$

$$= \left[\frac{7.08171282}{2.333405788} \right] (0.86711)$$

$$= 3.034925539\,(0.86711)$$

$$r \neq 2.631638777$$

EQUATION (21)

$$r = \left[\frac{4.9374\,(1.4343)}{1.41068\,(1.6541)} \right] \div 0.86711$$

$$= \left[\frac{7.08171282}{2.333405788} \right] \div 0.86711$$

$$= 3.034925539 \div 0.86711 \qquad r = 3.500014178$$

$$m_o \, v_o \sin\gamma_o = m_f \, v_f \sin\gamma_f \quad [8] \quad (EQ.\ 22)$$

$$[4.9374(1.4347)]\sin 14.28 = [1.41068(1.6541)]\sin 59.72$$

$$(7.08368778)\sin 14.28 = (2.333363468)\sin 59.72$$

$$1.74726778 \neq 2.01502645$$

The above formula [8] pertains to the Axial Flow of propellants from the impingement point and like the formula for the injector alignment, it fails to compute and must be restructured to equate mathematically.

This second statement of equalization determines if the Axial Flow is aligned along the axis of the combustion chamber while having a tan δ value equaling zero.

Inserting a multiplication sign for a equal sign in EQ. 22 yields an incorrect value since the tan δ value equals zero.

Therefore, a division sign is inserted for the equal sign which yields a value <1 but, not quite zero as it should if the Axial Flow is to equal a tan δ of zero, as according to the Sutton & Biblarz concept.

The calculated quotient of this modified formula has a value close to the value of the Velocity Ratio (V_r) as shown.

$$\tan \delta = Vr - \left[(m_o v_o \sin\gamma_o) / (m_f v_f \sin\gamma_f)\right] \quad \text{(EQ. 23)}$$
$$\tan \delta = 0$$

$$\tan \delta = 0.867118 - \frac{\left[4.9374(1.4343) \sin 14.28\right]}{\left[1.41068(1.6541) \sin 59.72\right]} = 0$$

$$= 0.867118 - \frac{1.746780577}{2.015062996} = 0$$

$$= 0.867118 - 0.8668615 = 0$$

$$\tan \delta = 0.000256548$$

$$\tan \delta = 0.0003 \cong 0$$

The algebraic purpose of subtracting from the Velocity Ratio value is clearly evident when considering that the concept for Axial Flow is to be equal to a tan of zero.

When subtracting the mass flows division from the Velocity Ratio value there is a remainder that nearly equals zero which confirms that the oxidant and fuel mixes and impinges in a axial flow along the combustion chamber axis for optimal propellant combustion.

In closing this paper, an explaination is required for the reasoning on how to the use the velocity ratio (Vr) for calculating the impingement angles with it's relation to the mix ratio and how the propellants flow with a tan of zero.

Mathematics has many ratios such as the Lift-Drag Ratio and π which dates back to the times of the Old Testament. In this paper, the Coefficient of Jet Contraction (Cc) is also defined as a ratio as are the trigonometric functions used.

Therefore, the idea of creating a Velocity Ratio (Vr) stems from the above history and was a thought of by this author then considered using it as a means to restructure the Sutton & Biblarz formulas to allow their impingement injection concept to equate correctly with mathematics.

Since the division of mass flows for oxidant and fuel equals the intended mix ratio; their velocities were selected to create this ratio then used in restructuring the propellant formulas so they equate correctly as Sutton & Biblarz did postulate for their optimal propellant combustion concept.

In an attempt to verify the restructured formulas' use for a Mix Ratio of 5 in a LOX/H mixture, the division for the sins of the injection angles proved not to correctly equate with this selected mix ratio. Although, this mix ratio did calculate correctly with Equation 21; where the quotient of the propellant mass flows times their velocities is divided by the velocity ratio, Vr; the division of the injection angle sin failed to do so.

Therefore, additional consideration is required to have the division of the injection angle sins equals correctly with the selected mix ratio of 5; since the mathematical result has to account for the additional oxidant of 30%.

Previously, this author stated that the impingement angle from Fig. 8-7 by Sutton & Biblarz resembles a 90 degree angle but, was proven to be less than 90 degrees with a cot value that equals the mix ratio of the propellants.

Yet, for a mix ratio of 5; the results has an incorrect value when dividing the injection angle sins for the mix ratio. This means that the determined impingement angle will be followed as previously outlined but, the value will be utilized differently when calculating the propellant injection angles.

For the 5 Mix Ratio, the impingement angle value as calculated with the 3.5 mix process will then be deducted from 180 Degrees with the resembled 90 Degree impingement angle from FIG. 8-7 where the difference would be the oxidant injection angle and this impingement angle would also serve as the fuel injection angle.

$$\theta_o = 180 - (Imp.\ \theta + \theta_1) \qquad 180 = 90 + (\theta_o + \theta_1)$$

$$= 180 - (90 + 78.69) \qquad = 90 + (11.31 + 78.6)$$

$$= 180 - 168.69 \qquad = 90 + 90$$

$$\theta_o = 11.31° \qquad 180 = 180$$

This diagram illustrates the propellant injection angles.

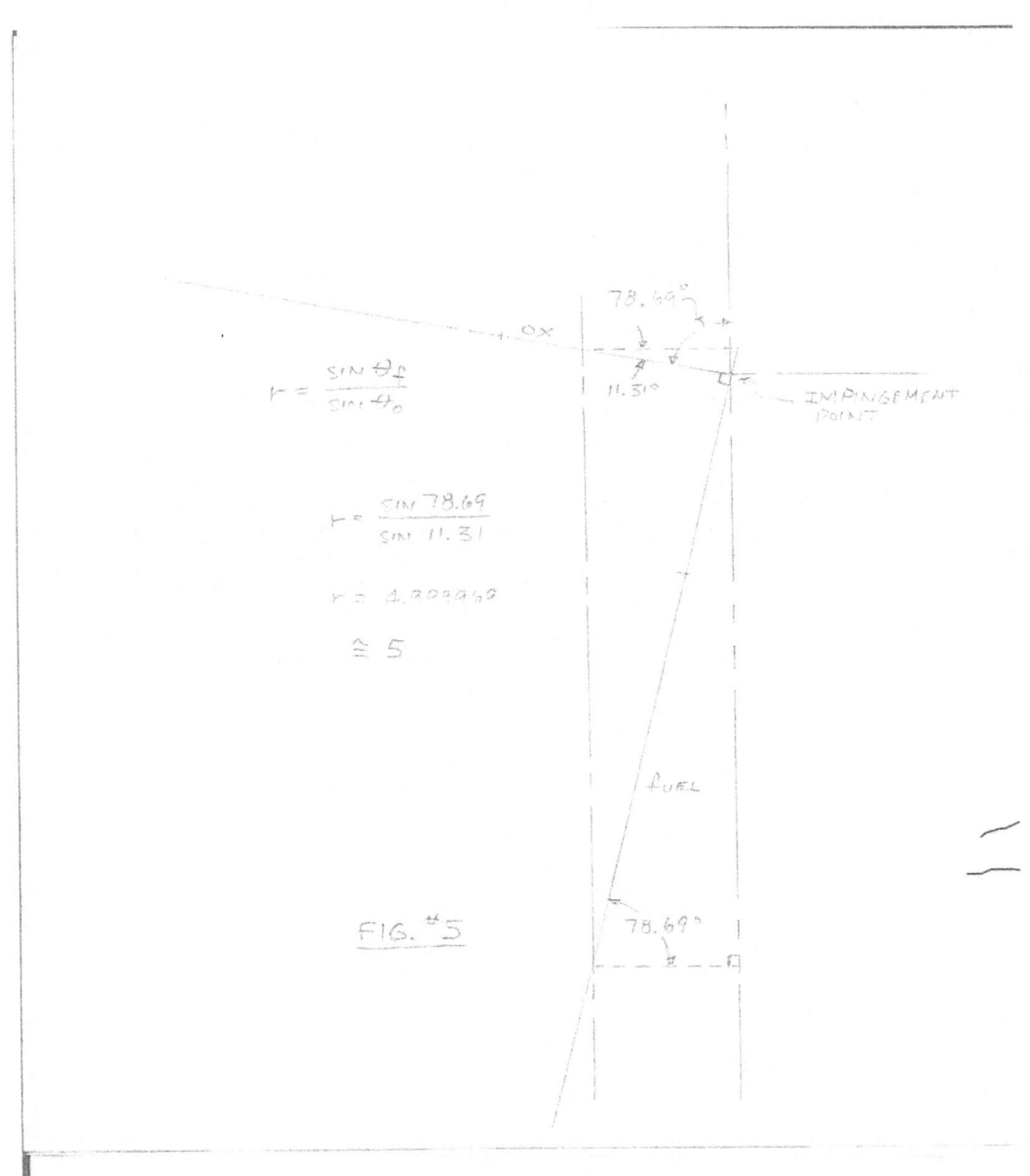

$$r = \frac{\sin \theta_f}{\sin \theta_o}$$

$$r = \frac{\sin 78.69}{\sin 11.31}$$

$$r = 4.999949$$

$$\cong 5$$

FIG. #5

A method to reconcile the results of the oxidant injection angles is to plot them on a chart where the increasing injection angle is along the 'x' axis while the mix ratio values increase on the 'y' axis. This would be best done with 10 sqaure grid paper since the values being worked with are in decimal units.

A line that connects these points would be the form of confirmation for having an additional 30% oxidant where the slope of 30 degrees correlates with this additional oxidant for the combustion of propellants in a 5Mix Ratio.

Another chart comparison is creating a right triangle where the units of difference for the injection angles and the mix ratios are plotted with the connecting line measured for 30 degrees to correlate with the 30 % additional oxidant.

The next pair of figures will illustrate these mathematics

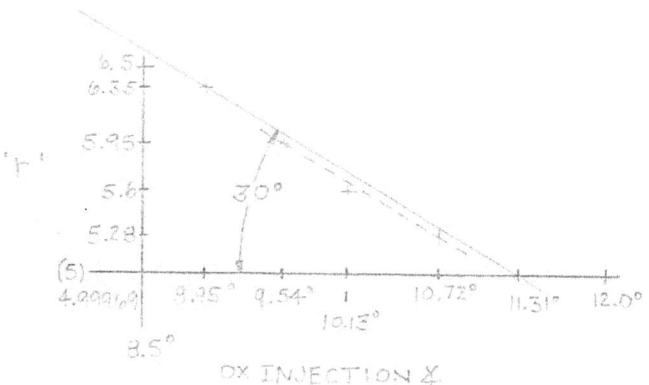

FIG. #6A

$$\tan \theta = \frac{1.35}{2.36}$$
$$= 0.5720338$$
$$\theta = 29.77°$$

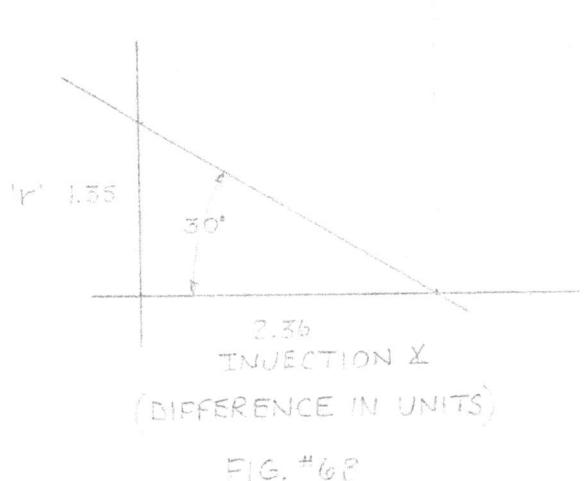

'r' 1.35

30°

2.36

INJECTION X

(DIFFERENCE IN UNITS)

FIG. #6P

The previous mathematical work pertained to one pair of propellants using a single mix ratio. However, to validate the functionality for other propellant pairs and mix ratios additional work is required to prove the premise of these restructured formulas stated as in Equations #21 & #22.

What follows is the validation of the previous math work where the results are presented in tabular form for these propellant pairs and mix ratios which are: LOX/RP-1, LOX/N3H and where Nitrogen Tetroxide is the oxidant and the fuel being Liquid Methane.

There is one final topic before concluding this work and this concerns the angles of propellant injection. A more concise and cohesive mathematical structure will be outlined to determine the propellant injection angles being equal to the mix ratio by the division of their sin values and their results will also be presented in tabular form.

Balanced Injector Orientation
for Propellant Combustion
Elmer A. Beck

TABLE #1

THRUST: 74200 LBs

Propellant Pair	Density (slugs/ft3)	Mix Ratio (r)	I sp
ox Nitrogen Tetroxide	1.067895		
		4.27	273
f Liquid Methane	0.825686		
ox LOX	0.00262		
		2.29	289
f RP-1	1.56		
ox LOX	0.00262		
		1.28	287
f N3H	0.00141		

Balanced Injector Orientation
for Propellant Combustion
Elmer A. Beck

TABLE #2

THRUST: 74200 Lbs	Propellant Pair (ox/f)	
Equation	LOX/RP-1	LOX/N3H
$C = Isp\,(g)$	9305.8	9241.4
$q = F/C$	7.97352189	8.0290865
W_o Eq. 4	5.549959	4.5075573
W_f Eq. 5	2.42356288	3.521529168
γ_o	23.6	38
γ_f	66.4	52
F_o Eq. 12	0.618369135	0.50572595
F_f Eq. 12	0.27003019	0.39509839
$A = \rho/Force$		
ox	0.00423695	0.00518067
f	5.7771318	0.00356873
Dia.		
ox	0.0938	0.0938
f	2.7188	0.0625

Balanced Injector Orientation
for Propellant Combustion
Elmer A. Beck

TABLE #2 (cont'd)

THRUST: 74200 Lbs Equation	Propellant Pair (ox/f)	
	LOX/RP-1	LOX/N3H
ox A_D	0.441786467	0.441786467
ox A_d	0.006902914	0.006902914
f A_D	9.6211275	0.441786467
f A_d	5.80535029	0.003067962
C_c Eq. 15		
ox	0.015625001	0.0015625
f	0.6033961	0.00694444
C_d Eq. 16		
ox	0.015625	0.015625
f	0.647862	0.00694444
ΔP Eq.-17 X: 18		
ox ρ	0.663439	1.18642285
f ρ	2069.57709	1.04601438
V_o Eq. 18	1.995321456	2.668284
V_f Eq. 18	189.366858	1.51788785
$V_r = V_o / V_f$	0.010536804	1.7578928

TABLE #2 (cont'd)

Verifying the mix ratio with the quotient of the division of the mass flows times their velocity then confirming that the axial flow will equal zero.

$$r = \left[\frac{m_o\, v_o}{m_f\, v_f} \right] \div v_r \qquad\qquad (EQ.\ 21)$$

$$r = \left[\frac{6.839158(65.13555)}{1.6016763(52.09092)} \right] \div 1.25042$$

$$= \left[\frac{445.4723845}{83.43279937} \right] \div 1.25042$$

$$r = 5.339295671 \div 1.25042$$

$$= 4.27$$

TABLE #2 (cont'd)

$$\tan \delta = 0 = v_r - \left[(m_o v_o \sin \gamma_o) / (m_f v_f \sin \gamma_f) \right]$$

(EQ. 22)

$$\tan \delta = 1.25042 - \frac{\left[(6.839158(65.13555) \sin 13.18 \right]}{\left[(1.6016763(52.09092) \sin 76.82 \right]}$$

$$= 1.25042 - \frac{101.5726097}{81.23505871}$$

$$= 1.25042 - 1.250354358$$

$$\tan \delta = 0.000066174$$

$$\tan \delta \cong 0$$

This method for the design of the injection angles is rooted in the relationships of the trigonometric functions; where the mix ratio will serve as the cot is for the impingement angle as stipulated at the start of this work for the design of impingement injectors.

After using the mix ratio for the cot of the impingement angle, it's angular value will be the fuel injection angle while it's sin value will be the cos for the oxidant injection value.

These angles should equal 90 degrees which is added with 90 degrees to equal 180 degrees since this would be the impingement plane as described previously in this text.

As an algebraic means to verify the accuracy of the propellant injection angles, the division of the sins for the propellant injection angles will have the quotient equaling the mix ratio of the propellants.

The equation to balance the mix ratio with the propellant injection angles will have the sin of the fuel injection angle divided by the sin of the oxidant injection angle.

This concludes a new means to determine and confirm the injection angles of bi-propellant combustion within a restructured mathematical approach for the formulas created by Sutton & Biblarz.

Balanced Injector Orientation
for Propellant Combustion
Elmer A. Beck

TABLE #3

(ox / f) MIX $r : \cot = 0.00 = \theta : \gamma_f$ $r = \sin \gamma_f / \sin \gamma_o$

RATIO $\sin \gamma_f : 0.00000 = \cos \gamma_o$ $[\gamma_f + \gamma_o + 90 = 180]$

LOX/H2

2.0
$r : \cot: 2.0 = 63.4$
$\sin 63.4: 0.89389 = \cos 26.6$
$r = \sin 63.4 / \sin 26.6$
$= 1.9995$
$\cong 2.0$

3.0
$r : \cot: 3.0 = 71.6$
$\sin 71.6: 0.948712 = \cos 18.4$
$r = \sin 71.6 / \sin 18.4$
$= 3.00086$
$\cong 3.0$

4.0
$r : \cot: 4.0 = 75.96$
$\sin 75.96: 0.97014 = \cos 14.04$
$r = \sin 75.96 / \sin 14.04$
$= 3.99888$
$\cong 4.0$

6.0
$r : \cot 6.0 = 80.54$
$\sin 80.54: 0.9862856 = \cos 9.5$
$r = \sin 80.5 / \sin 9.5$
$= 5.975$
$\cong 6.0$

7.5
$r : \cot 7.5 = 82.4$
$\sin 82.4: 0.9912386 = \cos 7.59$
$r = \sin 82.4 / \sin 7.59$
$= 7.5046$
$\cong 7.5$

LOX/ RP-1

1.5
$r : \cot 1.5 = 56.31$
$\sin 56.31: 0.83205 = \cos 33.69$
$r = \sin 56.31 / \sin 33.69$
$= 1.500004$

2.3
$r : \cot 2.3 = 66.5$
$\sin 66.5: 0.91706 = \cos 23.5$
$r = \sin 66.5 / \sin 23.5$
$= 2.2998$
$\cong 2.3$

Balanced Injector Orientation
for Propellant Combustion
Elmer A. Beck

TABLE #3 (cont'd)

(ox / f)	MIX	$r : \cot = 0.00 = \theta : \gamma_f$	$r = \sin \gamma_f / \sin \gamma_o$
	RATIO	$\sin \gamma_f : 0.00000 = \cos \gamma_o$	$[\gamma_f + \gamma_o + 90 = 180]$

LOX/ RP-1

$r : \cot = 3.0 = 71.57$

3.0 $\sin 71.57: 0.94871 = \cos 18.43$ $r = \sin 71.57/\sin 18.43$
$= 3.000864$

$r : \cot = 3.5 = 74.05$

3.5 $\sin 74.05: 0.9615 = \cos 15.95$ $r = \sin 74.05/\sin 15.95$
$= 3.4989$
$\cong 3.5$

LOX/ N3H

$r : \cot = 1.28 = 52$

1.28 $\sin 52: 0.788024 = \cos 38$ $r = \sin 52/ \sin 38$
$= 1.27994$
$\cong 1.28$

$r: \cot 4.27 = 76.82$

Nitrogen Tetroxide (ox)   4.27 $\sin 76.82: 0.973659 = \cos 13.18$ $r = \sin 76.82/\sin 13.18$
$= 4.27022$
$\cong 4.27$

Liquid Methane (f)

Balanced Injector Orientation
for Propellant Combustion
Elmer A. Beck

-REFERENCES-

[1] George P. Sutton & Oscar Biblarz; Thrust Chambers, Rocket Propulsion Elements 8th Ed.; Wiley 2011, p. 284

[2] George P. Sutton & Oscar Biblarz; Thrust Chambers, Rocket Propulsion Elements 8th Ed.; Wiley 2011, p. 283 FIG. 8-7

[3] Hunter Rouse; Appendix, Elementary Mechanics of Fluids; John Wiley & Sons, 1946; p.357

[4] Hunter Rouse; Appendix, Table VIII, Elementary Mechanics of Fluids; John Wiley & Sons, 1946; p.358

[5] NASA Design Drawing of Nozzle Injectors, unknown Report number

[6] Hunter Rouse; Pressure Variation in Accelerated Flow, Elementary Mechanics of Fluids; John Wiley & Sons, 1946, p. 56

[7] Hunter Rouse; Pressure Variation in Accelerated Flow, Elementary Mechanics of Fluids; John Wiley & Sons, 1946, p. 59

[8] George P. Sutton & Oscar Biblarz; Thrust Chambers, Rocket Propulsion Elements 8th Ed.; Wiley 2011 p.283 Eq. 8-7